I0446578

QUANTUM PHYSICS

A Beginner's Guide To Particle Physics - Navigating The Strange And Fascinating World Of Quantum Physics & More

JASE ROBBIN

Contents

Introductory

Quantum physics, also known as quantum mechanics, addresses the investigation of the characteristics and interrelationships of subatomic particles and atoms.

This particular branch of physics is indispensable for the comprehension of the physical universe. It provides a conceptual framework for understanding the characteristics and behaviors of particles at the quantum level, where the principles of classical physics cease to apply.

Quantum physics is comprised of the following fundamental concepts and principles:

• The phenomenon of wave-particle duality is observed in quantum particles, including electrons and photons, which manifest characteristics of both waves and particles. This duality presents a challenge to the traditional concept that particles follow unique classical trajectories.

• Energy Quantization: Energy is expressed in discrete units referred to as quanta. This is demonstrated by phenomena such as the photoelectric effect and quantized energy levels in atoms.

• Quantum Superposition: The simultaneous existence of multiple states is a property of quantum systems referred to as superposition. This is frequently demonstrated through the renowned thought experiment known as Schrodinger's cat.

• Entanglement: A phenomenon known as quantum entanglement occurs when particles become interconnected to the extent that their states are instantaneously correlated, despite the vast distances separating them.

• Uncertainty Principle: Position and momentum are two instances of pairings of attributes that, in accordance with this concept proposed by Werner Heisenberg, cannot be simultaneously known with absolute precision. As the precision of one property increases, the precision of the other property decreases.

• Quantum Tunneling: Particles have the ability to traverse energy barriers that are seemingly impenetrable according to classical physics. This phenomenon possesses ramifications across multiple disciplines, such as nuclear physics and electronics.

- Wavefunctions and Quantum States: The quantum system's state is characterized by a mathematical construct known as the wavefunction. The wavefunction provides data regarding the probabilities associated with locating a particle in various states.

Technological progress has resulted from quantum physics, including the creation of semiconductors, lasers, and specific medical imaging techniques. Furthermore, it serves as the foundation for quantum computing and carries significant ramifications for our fundamental comprehension of the essence of reality. Complex and frequently counterintuitive, the discipline challenges our traditional understanding of the physical world.

Wave-Particle Duality

Wave-particle duality is a fundamental concept in quantum physics that describes the dual nature of particles at the quantum level. It suggests that particles, such as electrons and photons, can exhibit both wave-like and particle-like properties, depending on the experimental conditions. This phenomenon challenges the classical Newtonian view of particles as distinct, localized entities with well-defined trajectories.

The key points of wave-particle duality include:

Wave-Like Properties:

- **Interference:** Like waves, particles exhibit interference patterns when overlapping waves combine constructively or destructively. This phenomenon is observed in experiments such as the double-slit experiment.

• **Diffraction:** Particles can undergo diffraction, a wave-like behavior where they bend around obstacles or openings. This is also evident in the double-slit experiment.

Particle-Like Properties:

• **Photons:** Light, which was traditionally considered a wave, was found to exhibit particle-like behavior in certain experiments. The photoelectric effect, explained by Einstein in 1905, demonstrated that light could be described as discrete packets of energy called photons.

• **De Broglie's Hypothesis:** Louis de Broglie proposed in 1924 that particles, not just photons, also exhibit wave-like characteristics. He suggested that the wavelength of a particle is inversely proportional to its momentum, introducing the concept of matter waves.

Double-Slit Experiment:

• The double-slit experiment is a classic illustration of wave-particle duality. When particles such as electrons or photons are directed through two closely spaced slits onto a screen, an interference pattern emerges, suggesting a wave-like nature. However, when the particles are observed, they behave like individual particles, creating a pattern consistent with particle trajectories.

Quantum Superposition:

• Wave-particle duality is closely related to the idea of quantum superposition. In a superposition state, a particle can exist in multiple states simultaneously, as described by its wavefunction. This concept is fundamental to understanding the behavior of particles at the quantum level.

Observational Effects:

• The act of measurement or observation collapses the

wavefunction, forcing the particle to "choose" a specific state. This is often referred to as the collapse of the wavefunction.

Wave-particle duality is a central aspect of quantum mechanics and is crucial for understanding the behavior of matter and energy at the microscopic scale. It highlights the inherently probabilistic nature of quantum systems, where predicting the precise behavior of particles is constrained by the uncertainty principle. This duality is one of the key features that sets quantum mechanics apart from classical physics.

CHAPTER TWO
Quantum Mechanics Basics

Quantum mechanics is a branch of physics that describes the behavior of matter and energy at the smallest scales, typically at the level of atoms and subatomic particles. Here are some key concepts and principles that form the basics of quantum mechanics:

Wavefunctions and Probability:

• **Wavefunction** (ψ): The wavefunction is a mathematical function that describes the quantum state of a system. It contains information about the probabilities of finding a particle in different positions and states.

• **Probability Density:** The square of the absolute value of the wavefunction ($|\psi|^2$) gives the probability density of finding a particle in a particular location. Regions with higher probability density are more likely locations for the particle.

Quantization of Energy:

• **Quantum States and Energy Levels:** In quantum mechanics, energy levels are quantized, meaning they come in discrete, distinct values. This is evident in the electron energy levels within atoms.

Quantum Superposition:

• **Superposition Principle:** Quantum systems can exist in multiple states simultaneously. This is known as superposition. For example, an electron can exist in a superposition of different energy states.

Quantum Entanglement:

• **Entanglement:** Quantum entanglement occurs when two or more particles become correlated in such a way that the state of one particle instantaneously influences the state of another, even if they are far apart. This phenomenon is often described as "spooky action at a distance."

Uncertainty Principle:

• **Heisenberg's Uncertainty Principle:** Formulated by Werner Heisenberg, this principle states that certain pairs of properties, such as position and momentum, cannot both be precisely known simultaneously. The more precisely one property is known, the less precisely the other can be known.

Quantum Tunneling:

• **Quantum Tunneling:** Particles can "tunnel" through energy barriers that classical physics would suggest are impassable. This phenomenon has implications in various fields, including electronics and nuclear physics.

Wave-Particle Duality:

• **Wave-Particle Duality:** Particles, such as electrons and photons, exhibit both wave-like and particle-like properties. This duality is a fundamental aspect of quantum mechanics and is illustrated in experiments like the double-slit experiment.

Quantum Measurement:

- **Wavefunction Collapse:** When a measurement is made on a quantum system, the system's wavefunction collapses to one of its possible states. The outcome of the measurement is probabilistic, and the act of measurement is a central concept in understanding quantum mechanics.

Quantum Operators:

- **Operators:** In quantum mechanics, physical properties (like position, momentum, and energy) are associated with mathematical operators. These operators act on the wavefunction to extract information about the system.

Quantum States and Observables:

• **Eigenstates and Eigenvalues:** A quantum state can be an eigenstate of a particular observable, and the corresponding eigenvalue represents the value that would be measured if the property associated with the observable is measured.

These basic principles of quantum mechanics lay the foundation for a deeper understanding of the behavior of matter and energy at the quantum level. The theory has been remarkably successful in explaining a wide range of phenomena and is an essential framework for modern physics.

Quantum Superposition

One of the cornerstones of quantum physics is the idea of quantum superposition, which claims that quantum systems can exist in more than one state or configuration at the same time. A defining characteristic of quantum mechanics that sets it apart from conventional physics is this concept, which questions classical intuitions.

Here are the key aspects of quantum superposition:

States and Wavefunctions:

• In quantum mechanics, the state of a system is described by a mathematical entity called a wavefunction (often denoted by the symbol ψ).

• The wavefunction contains information about the probabilities of finding a particle in different states or configurations.

Linear Combination of States:

• A superposition is a linear combination of two or more

quantum states. Mathematically, if $|A\rangle$ and $|B\rangle$ are two possible states of a system, then a superposition state $|\psi\rangle$ can be expressed as:

$$|\psi\rangle = c_A|A\rangle + c_B|B\rangle |\psi\rangle = c_A|A\rangle + c_B|B\rangle$$

• Here, $c_A c_A$ and $c_B c_B$ are complex numbers known as probability amplitudes, and their squares give the probabilities of finding the system in the corresponding states.

Quantum Interference:

• Superposition leads to the phenomenon of quantum interference, where the probability amplitudes of different states can interfere constructively or destructively.

• Constructive interference increases the probability of finding the system in certain states, while destructive interference decreases it.

Example: Double-Slit Experiment:

• The double-slit experiment is a classic example illustrating superposition. When particles, such as electrons or photons, are sent through two slits, the resulting interference pattern on a screen suggests that each particle took multiple paths and interfered with itself.

Measurement and Collapse:

• When a measurement is made on a quantum system, the superposition collapses to one of the possible states, and the system is found in a definite state.

• The act of measurement forces the system to "choose" one of its possible states, with the probability of each outcome determined by the squared magnitudes of the probability amplitudes.

Entanglement and Superposition:

• Superposition is closely related to the concept of entanglement. In an entangled state, the properties of two or more particles are correlated, and the system as a whole exists in a superposition of entangled states.

The evolution of quantum technologies, including quantum computing, is greatly influenced by quantum superposition, which is an essential part of quantum physics. It shows that classical intuitions can't explain how particles behave at the quantum level and that quantum systems are inherently probabilistic.

CHAPTER THREE
Quantum Measurement And Uncertainty

When it comes to understanding how particles behave at the quantum level, two interrelated ideas in quantum mechanics that greatly impact our knowledge are quantum measurement and uncertainty.

Quantum Measurement:

• In quantum mechanics, when a measurement is made on a quantum system, the act of measurement causes the system to "collapse" into one of its possible states. This collapse is a sudden and nondeterministic transition from a superposition of states to a definite state.

• The outcome of a measurement is probabilistic, and the probabilities are determined by the squared magnitudes of the probability amplitudes associated with each possible state in the system's superposition.

• The process of measurement is often described by the Born rule, which provides a mathematical formula for calculating the probabilities of different outcomes based on the wavefunction.

Wavefunction Collapse:

• The wavefunction, which evolves deterministically according to the Schrödinger equation between measurements, undergoes a sudden collapse when a measurement is made. After the collapse, the system is found in one of its eigenstates.

• The nature of wavefunction collapse is still a topic of philosophical and interpretational debates in quantum mechanics.

Uncertainty Principle:

• Formulated by Werner Heisenberg, the uncertainty principle is a fundamental concept in quantum mechanics that sets limits on the precision with which certain pairs of properties of a particle can be simultaneously known.

• The most well-known form is the position-momentum uncertainty principle: $\Delta x \cdot \Delta p \geq \hbar 2 \Delta x \cdot \Delta p \geq 2\hbar$

• Here, Δx is the uncertainty in position, Δp is the uncertainty in momentum, and $\hbar$ (h-bar) is the reduced Planck constant.

Implications of the Uncertainty Principle:

• The uncertainty principle implies that there are inherent limits to the precision with which we can simultaneously measure certain pairs of conjugate variables, such as position and momentum or energy and time.

• The more accurately we know one of these variables, the less accurately we can know the other. This is not due to limitations in measurement devices but is a fundamental property of quantum systems.

Wave-Particle Duality and Measurement:

• The uncertainty principle is closely related to the wave-particle duality of quantum particles. The more localized a particle's wavefunction is in space (i.e., the more accurately we know its position), the more spread out its possible momenta become, and vice versa.

Both quantum measurement and the uncertainty principle highlight the probabilistic and non-intuitive nature of quantum mechanics. They challenge classical intuitions based on deterministic laws and continuous trajectories and are essential for understanding the behavior of particles at the quantum level.

These principles have practical implications for technologies like quantum computing and are foundational to the entire framework of quantum mechanics.

Quantum States And Operators

The idea of quantum states and operators is fundamental to comprehending how quantum systems behave in quantum mechanics. The following is a synopsis of these core ideas:

Quantum States:

State Vector/Wavefunction (ψ):

• The state of a quantum system is described by a mathematical entity known as a wavefunction, often denoted by the symbol ψ (psi).

• The wavefunction contains information about the probabilities of finding a particle in different states or configurations.

• The square of the absolute value of the wavefunction ($|\psi|^2$) gives the probability density of finding the particle in a particular state.

Superposition:

• Quantum systems can exist in multiple states simultaneously, a concept known as superposition. A superposition state is a linear combination of two or more basis states.

• Mathematically, if $|A\rangle$ and $|B\rangle$ are two possible states, a superposition state $|\psi\rangle$ can be expressed as: $|\psi\rangle=c_A|A\rangle+c_B|B\rangle|\psi\rangle=c_A|A\rangle+c_B|B\rangle$, where $c_A c_A$ and $c_B c_B$ are complex numbers.

Quantum Measurement:

• When a measurement is made on a quantum system, the system's wavefunction collapses to one of the possible eigenstates of the measured observable.

• The probabilities of different outcomes are determined by the squared magnitudes of the probability amplitudes associated with each state.

Operators:

• In quantum mechanics, physical observables are associated with mathematical operators. These operators act on the wavefunction to extract information about the system's properties.

• For example, the position operator (XX) corresponds to the position of a particle, and the momentum operator (PP) corresponds to its momentum.

Eigenstates and Eigenvalues:

• An eigenstate of an operator is a state for which the action of the operator is equivalent to multiplication by a constant (the eigenvalue). Mathematically, $A|a\rangle = a|a\rangle A|a\rangle = a|a\rangle$, where AA is the operator, $|a\rangle|a\rangle$ is the eigenstate, and aa is the eigenvalue.

• Observables in quantum mechanics are associated with Hermitian operators, and the possible outcomes of measurements are the eigenvalues of those operators.

Commutation Relations:

• The behavior of quantum operators can be described by commutation relations. The commutator of two operators A and B is given by $[A,B]=AB-BA$.

• Commutation relations provide information about the compatibility of measurements and the inherent uncertainties associated with certain pairs of observables.

Time Evolution Operator:

• The time evolution of a quantum system is described by the time evolution operator, often denoted as

U(t)U(t). This operator evolves the state of the system from one instant in time to another.

• The Schrödinger equation governs the time evolution of a quantum system: $i\hbar \frac{d}{dt}|\psi\rangle = H|\psi\rangle i\hbar \frac{d}{dt}|\psi\rangle = H|\psi\rangle$, where HH is the Hamiltonian operator.

Understanding quantum states and operators is essential for performing calculations in quantum mechanics, predicting the outcomes of measurements, and developing a comprehensive understanding of the quantum behavior of particles.

The mathematical formalism involving operators and wavefunctions is a powerful tool for representing and manipulating quantum information.

CHAPTER FOUR
Quantum Tunneling And Barrier Penetration

When particles engage in quantum tunneling, they are able to traverse energy barriers that, by the rules of conventional physics, ought to be insurmountable. This peculiar behavior is a hallmark of quantum physics and is caused by the wave-like quantum state of particles.

The main features of quantum tunneling and how they penetrate barriers are as follows:

Wave-Particle Duality:

• Particles at the quantum level, such as electrons, exhibit both wave-like and particle-like properties. The wave aspect allows particles to spread out in space, leading to phenomena like interference and tunneling.

Barrier Potential:

• In classical physics, a particle with insufficient energy cannot overcome a potential energy barrier. However, in quantum mechanics, there is a non-zero probability that the particle can penetrate the barrier and appear on the other side, even if its energy is lower than the barrier height.

Probability Amplitude and Tunneling Probability:

• The probability of a particle tunneling through a barrier is determined by the probability amplitude, which is related to the wavefunction of the particle.

• The tunneling probability decreases exponentially with the width and height of the barrier. The narrower and taller the barrier, the less likely the particle is to tunnel through.

Quantum Mechanical Tunneling Equation:

• The tunneling probability (TT) can be calculated using the quantum mechanical tunneling equation. For a one-dimensional barrier, the transmission coefficient (TT) is given by: $T=e-2\alpha dT=e-2\alpha d$ where $\alpha\alpha$ is the tunneling constant, dd is the width of the barrier, and ee is the base of the natural logarithm.

Applications of Quantum Tunneling:

• **Nuclear Fusion:** Quantum tunneling is crucial in explaining certain nuclear fusion reactions in stars and in experimental fusion research.

• **Scanning Tunneling Microscopy (STM):** Tunneling is utilized in STM to image surfaces at the atomic level.

• **Semiconductor Devices:** Quantum tunneling is a key mechanism in the operation of tunnel diodes and quantum tunneling devices in semiconductor physics.

Quantum Tunneling in Alpha Decay:

• In nuclear physics, alpha decay involves the quantum tunneling of an alpha particle through a potential barrier created by the nuclear forces holding the alpha particle within the nucleus.

Tunneling Time:

• The concept of tunneling time is the time it takes for a particle to traverse a barrier. However, the definition and measurement of tunneling time are subjects of ongoing debate and research in the field of quantum mechanics.

An intriguing and paradoxical feature of quantum mechanics, quantum tunneling has been verified experimentally in a number of different situations. Along with challenging our traditional intuitions about particle behavior and the existence of barriers, it has major ramifications in varied domains such as nuclear physics and electronics.

Quantum Angular Momentum

When particles are in motion, one of their quantum mechanical properties is their angular momentum. A vector quantity associated with an object's rotational motion, angular momentum is defined in classical physics.

Angular momentum can only take on specific discrete values in quantum physics since it is quantized.

Here are the key aspects of quantum angular momentum:

Quantization of Angular Momentum:

• In quantum mechanics, angular momentum is quantized in units of $\hbar\hbar$ (the reduced Planck constant), denoted as l$\hbar$l$\hbar$, where ll is a quantum number.

• The quantization condition is given by L=l$\hbar$L=l$\hbar$, where LL is the total angular momentum and ll is an integer or half-integer known as the azimuthal quantum number.

Eigenstates of Angular Momentum:

• The angular momentum operator (LL) acts on quantum states, and the eigenstates of LL are associated with specific values of angular momentum.

• The eigenvalues of the squared angular momentum operator (L2L2) are given by $l(l+1)\hbar2l(l+1)\hbar2$, where ll is the azimuthal quantum number.

Angular Momentum Operators:

• There are two components of the angular momentum operator: $L_z L_z$, which measures angular momentum along the z-axis, and $L_2 L_2$, which measures the total squared angular momentum.

• The operators $L_x L_x$, $L_y L_y$, and $L_z L_z$ do not commute, reflecting the noncommutative nature of quantum observables.

Quantum Numbers:

• In addition to the azimuthal quantum number (ll), there is also the magnetic quantum number (mm) associated with the zzcomponent of angular momentum. The values of mm range from –l–l to ll in integer steps.

Angular Momentum States:

• The eigenstates of the angular momentum operators are spherical harmonics, and they describe the distribution of probability density for finding a particle with angular momentum in different directions.

Conservation of Angular Momentum:

• Similar to classical mechanics, the total angular momentum of an isolated system is conserved in quantum mechanics.

• Angular momentum conservation plays a crucial role in understanding phenomena such as the quantization of electron orbits in atoms.

Spin Angular Momentum:

• Particles, such as electrons, also possess an intrinsic angular momentum called spin. Spin is a fundamental property and is quantized in units of $\hbar/2\hbar/2$.

• The total angular momentum of a particle is the sum of its orbital angular momentum and its spin angular momentum.

Atomic structure, electron behavior in magnetic fields, and interpretation of angular momentum in quantum systems are all affected by quantum angular momentum, which is an important feature of the quantum description of particles. One of the basic ideas that separates classical physics from quantum mechanics is the quantization of angular momentum.

CHAPTER FIVE
Quantum Mechanics In Three Dimensions

Quantum mechanics in three dimensions expands on the principles of quantum mechanics to describe the behavior of particles in three-dimensional space.

The extension from one to three dimensions introduces additional complexity but follows the same fundamental principles. Here are some key aspects of quantum mechanics in three dimensions.

Wavefunction in Three Dimensions:

• In three-dimensional quantum mechanics, the wavefunction ($\psi\psi$) becomes a function of three spatial coordinates: $\psi(x,y,z)\psi(x,y,z)$.

• The wavefunction describes the probability amplitude of finding a particle at a particular position in three-dimensional space.

Schrödinger Equation in Three Dimensions:

• The Schrödinger equation for a particle in three dimensions is given by: $-\hbar 22m\nabla 2\psi + V(x,y,z)\psi = E\psi - 2m\hbar 2$ $\nabla 2\psi + V(x,y,z)\psi = E\psi$

- Here, $\nabla^2 \nabla^2$ is the Laplacian operator, and $V(x,y,z)V(x,y,z)$ represents the potential energy function.

Quantum States and Quantum Numbers:

- In addition to the azimuthal quantum number ($l$$l$) and the magnetic quantum number ($m$$m$) associated with angular momentum, three-dimensional quantum mechanics introduces the principal quantum number ($n$$n$) to describe the energy levels of a quantum system.

- The set of quantum numbers (n,l,m_n,l,m) fully characterizes a quantum state in three dimensions.

Angular Momentum in Three Dimensions:

- The angular momentum operators $(L_x, L_y, L_z L_x, L_y, L_z)$ have three components, and the eigenstates are associated with three quantum numbers: $l, m_l, m_s l, m_l, m_s$.

- The eigenvalues of the squared angular momentum operator $(L_2 L_2)$ are given by $l(l+1)\hbar_2 l(l+1)\hbar_2$.

Hydrogen Atom as a Model:

• The hydrogen atom is a key example of a system described by quantum mechanics in three dimensions. The solutions to the Schrödinger equation for the hydrogen atom yield quantized energy levels and specific angular momentum states.

Orbital Shapes:

• Three-dimensional quantum mechanics leads to a variety of orbital shapes, including spherically symmetric s orbitals, the dumbbell-shaped p orbitals, and more complex d and f orbitals.

Quantum Tunnelling in Three Dimensions:

• Quantum tunneling, the phenomenon where particles pass through potential barriers, can occur in three dimensions. The probability of tunneling is influenced by the width and height of the barrier.

Quantum Mechanical Operators in Three Dimensions:

• Operators representing physical observables, such as position $(x,y,z x,y,z)$, momentum $(px,py,pz px ,py,pz)$, and angular momentum $(Lx,Ly,Lz Lx,Ly,Lz)$, are extended to three dimensions.

Quantum mechanics in three dimensions provides a more comprehensive framework for understanding the behavior of particles in realistic physical environments.

It is a crucial aspect of describing the quantum world, from the structure of atoms and molecules to the behavior of particles in various potentials and fields. The mathematical formalism becomes more intricate, but the underlying principles remain consistent with those in one-dimensional quantum mechanics.

Quantum Information And Computing

Quantum information and quantum computing are interdisciplinary fields that leverage the principles of quantum mechanics to process and store information in ways that go beyond classical computing.

These fields explore the use of quantum bits (qubits) and quantum gates to perform computations, promising the potential to solve certain problems exponentially faster than classical computers.

Here are key aspects of quantum information and quantum computing:

Quantum Information:

Qubits:

• Quantum bits, or qubits, are the fundamental units of quantum information. Unlike classical bits, qubits can exist in multiple states simultaneously due to superposition.

Superposition:

• Qubits can exist in a superposition of states, allowing them to represent both 0 and 1 simultaneously. This property enables quantum computers to process information in parallel.

Entanglement:

• Entanglement is a quantum phenomenon where two or more qubits become correlated in such a way that the state of one qubit instantaneously influences the state of another, even when separated by large distances.

Quantum States and Quantum Gates:

• Quantum states are described using wavefunctions or density matrices. Quantum gates, analogous to classical logic gates, manipulate these states to perform quantum computations.

• Gates such as Hadamard gates, CNOT gates, and phase gates are fundamental in quantum circuits.

Quantum Computing:

Quantum Algorithms:

• Quantum algorithms are designed to leverage the unique properties of quantum systems to solve specific problems more efficiently than classical algorithms.

• Examples include Shor's algorithm for factoring large numbers exponentially faster and Grover's algorithm for searching unsorted databases.

Quantum Circuits:

• Quantum circuits are the analogs of classical circuits in quantum computing. They consist of qubits and quantum gates arranged in a sequence to perform a specific computation.

Quantum Error Correction:

• Quantum error correction is essential in quantum computing due to the inherent fragility of quantum states. Quantum error correction codes, like the surface code, protect quantum information from errors caused by decoherence and other disturbances.

Quantum Hardware:

• Quantum computers are built using various physical platforms, including superconducting qubits, trapped ions, and topological qubits. Companies and research institutions are actively developing and experimenting with quantum hardware.

Decoherence and Noise:

• Decoherence, caused by interactions with the environment, and quantum noise present challenges in maintaining the integrity of quantum information. Quantum error correction and fault-tolerant techniques are employed to address these issues.

Quantum Supremacy:

• Quantum supremacy refers to the point at which a quantum computer can perform a specific task more efficiently than the most powerful classical supercomputers. Achieving quantum supremacy is a significant milestone in the field.

Applications:

• Quantum computers have the potential to revolutionize various fields, including cryptography (breaking existing cryptographic schemes and creating quantum-safe encryption), optimization (solving complex optimization problems), and drug discovery (simulating molecular interactions).

Quantum information and quantum computing are active areas of research with the potential to bring about transformative changes in information processing.

However, the field is still in the early stages of development, and practical, large-scale quantum computers remain a significant technological challenge.

Researchers are actively working to overcome the technical hurdles and explore the full potential of quantum information processing.

CHAPTER SIX
Quantum Field Theory

Quantum Field Theory (QFT) is a theoretical framework that combines quantum mechanics and special relativity to describe the behavior of particles and fields at the most fundamental level. It provides a comprehensive and consistent framework for understanding the interactions of elementary particles and their dynamics. Here are key aspects of quantum field theory:

1. **Fields and Particles:**

• **Fields:** In QFT, physical quantities such as the electromagnetic field or the electron field are represented as quantum fields. These fields permeate all of spacetime.

• **Particles:** Particles are interpreted as excitations or quanta of these fields. For example, the photon is the quantum of the electromagnetic field, and the electron is the quantum of the electron field.

2. **Lagrangian Formalism:**

• QFT uses a Lagrangian formulation to describe the dynamics of fields. The Lagrangian density specifies the energy of the system as a function of field values and their derivatives.

3. **Quantization:**

• The process of quantization involves promoting classical fields to quantum operators. This results in the creation and annihilation operators that describe the creation and destruction of particles.

4. **Vacuum State:**

• The vacuum state in QFT is not an empty state but a state with the lowest possible energy. Virtual particle-antiparticle pairs continually pop in and out of existence, contributing to vacuum fluctuations.

5. Quantum Fields and Interactions:

• QFT describes interactions between particles as the exchange of virtual particles. Feynman diagrams are graphical representations used to calculate scattering amplitudes and describe these interactions.

6. Renormalization:

• Quantum field theories often involve infinities in their calculations. Renormalization is a process used to remove these infinities and obtain finite and physically meaningful results.

7. **Standard Model:**

• The Standard Model of particle physics is a quantum field theory that describes the electromagnetic, weak, and strong nuclear interactions. It includes fields for quarks, leptons, and gauge bosons.

8. **Spontaneous Symmetry Breaking:**

• Spontaneous symmetry breaking is a phenomenon in QFT where the ground state of a system does not exhibit the same symmetries as the underlying Lagrangian. This is responsible for giving mass to certain particles.

9. Quantum Chromodynamics (QCD):

• QCD is the quantum field theory describing the strong force that binds quarks and gluons into protons, neutrons, and other hadrons.

10. Quantum Gravity:

• While the Standard Model successfully describes three out of the four fundamental forces (excluding gravity), a consistent quantum field theory for gravity is still a topic of active research. String theory and loop quantum gravity are among the proposed approaches.

11. Quantum Field Theory in Curved Spacetime:

• QFT can be extended to curved spacetime, as described by general relativity. This is important for understanding the quantum aspects of black holes and the early universe.

Quantum Field Theory is a powerful and successful framework that has been instrumental in explaining a wide range of phenomena in particle physics. While the formalism is complex and the calculations can be challenging,

QFT has provided a deep and coherent understanding of the quantum nature of particles and their interactions.

Challenges In Unifying Quantum Mechanics And General Relativity

Achieving unification between general relativity and quantum mechanics is among the most formidable obstacles in theoretical physics.

These two seminal theories delineate the operations of the cosmos on extremely distinct scales: general relativity governs the macroscopic realm of spacetime and gravitation, whereas quantum mechanics governs the microscopic realm of particles. A number of obstacles manifest when endeavoring to integrate these theories into a solitary, coherent framework:

1. Quantum Gravity:

• The most central challenge is the quest for a theory of quantum gravity. General relativity successfully describes gravity as the curvature of spacetime caused by mass and energy, but it does not incorporate the principles of quantum mechanics.

Attempts to quantize gravity have encountered mathematical and conceptual obstacles.

Planck Scale Physics:

• The unification problem is often associated with the Planck scale, which represents the energy scale at which quantum gravitational effects become significant.

At this scale (around 10191019 GeV), both quantum and relativistic effects are expected to be important, and a theory of quantum gravity is needed.

Singularities and the Beginning of the Universe:

• Both general relativity and quantum mechanics predict singularities in certain conditions, such as at the center of black holes or at the beginning of the universe (the Big Bang). Resolving these singularities requires a consistent theory that combines quantum effects with the curved spacetime of general relativity.

Information Paradox in Black Holes:

• The study of black holes poses a significant challenge. The Hawking radiation predicted by quantum field theory suggests that black holes should emit radiation and eventually evaporate, violating the principle of information conservation. Resolving this paradox requires a quantum theory of gravity.

Nonlocality and Entanglement:

• Quantum mechanics involves nonlocal interactions and entanglement, which challenges our classical understanding of spacetime. Incorporating these features into a quantum theory of gravity requires a new conceptual framework.

Background Independence:

• General relativity is a background-independent theory, meaning the geometry of spacetime is dynamic and can evolve. Many quantum theories, however, rely on fixed background structures. Achieving a background-independent quantum theory of gravity is a theoretical challenge.

Mathematical Inconsistencies:

• Combining quantum mechanics and general relativity often leads to mathematical inconsistencies, such as infinities in calculations. Techniques like renormalization, which work well in quantum field theory, face challenges in the context of quantum gravity.

Emergent Spacetime:

• Some approaches suggest that spacetime itself may be an emergent concept from more fundamental degrees of freedom. Understanding how spacetime emerges from a more fundamental quantum structure is an open question.

Several theoretical frameworks have been proposed to address these challenges, including string theory, loop quantum gravity, and approaches based on noncommutative geometry. However, a conclusive and experimentally verified theory of quantum gravity remains elusive, and research in this area is ongoing.

The unification of quantum mechanics and general relativity represents a frontier in physics and is crucial for a complete understanding of the fundamental nature of the universe.

CHAPTER SEVEN
Practical Implementations Of Quantum Physics

Quantum physics is applicable to a vast array of disciplines, including healthcare and information technology. The following are noteworthy implementations of quantum physics:

• Quantum Computing: By utilizing the principles of quantum superposition and entanglement, quantum computers are capable of executing specific categories of computations at a significantly quicker rate than classical computers. Sub-quantum computation has the capacity to fundamentally transform domains including optimization, cryptography, and quantum system simulation.

• Quantum Cryptography: o Quantum key distribution (QKD) enables the exchange of cryptographic keys in complete secrecy by utilizing quantum principles to secure communication. It employs fundamental quantum mechanical principles to identify eavesdropping attempts.

• Quantum Sensors: Quantum sensors, including quantum

magnetometers and quantum gravimeters, enable precise measurements by capitalizing on quantum properties. The sensors find utility in the fields of medical imaging, navigation, and geophysics.

• Quantum Metrology: o By utilizing quantum properties, quantum metrology improves the accuracy of measurements. Quantum principles have practical implications in domains such as atomic devices, where they enhance the precision of timekeeping.

• Quantum Communication: Protocols for quantum

communication, such as quantum teleportation and quantum entanglement, facilitate the transmission of information in a secure and efficient manner. The aforementioned protocols are applicable to both quantum networks and quantum key distribution.

• Quantum imaging methods, such as quantum phantom imaging and quantum-enhanced imaging, provide imaging applications with improved resolution and sensitivity. The utilization of quantum imaging extends to microscopy, remote sensing, and medical imaging.

• The Utilization of Quantum Sensing in Medical Applications: In the realm of medicine, quantum sensors can be implemented to facilitate precise measurements, including the monitoring of molecular-level biological processes and the detection of minute magnetic fields in the brain (magnetoencephalography).

• Applications of Quantum Materials and Technologies: Quantum materials, such as quantum dots and superconductors, find utility in the fields of energy, sensors, and electronics. For instance, magnetic resonance imaging (MRI) devices utilize superconductors.

• Algorithms Inspired by Quantum Physics: Simple optimization challenges across industries have been resolved using quantum-inspired algorithms, including simulated annealing and quantum-inspired optimization, even prior to the widespread implementation of practical quantum computers.

• Quantum biology investigates the significance of quantum phenomena within the context of biological processes. This includes investigating the potential for quantum effects in brain processes and the investigation of quantum coherence in photosynthesis.

• The ongoing development of a quantum internet seeks to facilitate the transmission of quantum information over extensive distances in a secure and efficient manner. Quantum networks and quantum repeaters are fundamental elements of this nascent technology.

• Quantum Simulation: o Quantum simulators simulate the behavior of other quantum systems by utilizing controlled quantum systems. This has applications in materials science and chemistry, among others, for the comprehension of complex quantum phenomena.

Notwithstanding the fact that numerous quantum technologies are still in their nascent phases of development, continuous advancements and research broaden the scope of potential applications and operationalize quantum principles in a variety of fields.

Summary

Our knowledge of the universe's fundamental characteristics has been radicaly altered by quantum physics. The principles of quantum mechanics, which include quantum entanglement and wave-particle duality, challenge classical intuitions and offer a more precise characterization of the behavior of particles at the quantum level.

Quasi-mechanical principles have given rise to an array of revolutionary implementations, including quantum computing, quantum cryptography, sensors, and imaging technologies.

The utilization of quantum properties presents novel prospects in the domains of information processing, communication, and measurement, thereby potentially bringing about a paradigm shift in numerous sectors.

Nevertheless, considerable obstacles persist, most notably in the endeavor to integrate quantum mechanics and general relativity and construct a comprehensive theory of quantum gravity.

Ongoing inquiries in theoretical physics encompass, among others, the characteristics of singularities, the information paradox that arises in black holes, and the enigmatic pursuit of a quantum theory of gravity at the Planck scale.

An increasing number of insights continue to emerge from the interaction between quantum mechanics and other scientific disciplines, including materials science and biology, as we progress further into the quantum domain.

Once predominantly conceptual, quantum technologies are presently gaining traction in the real world and offer the potential for revolutionary progress in the fields of computation, communication, and sensing.

The exploration of the quantum realm is continuous, replete with profound revelations and enduring obstacles. In addition to being a potent instrument for technological advancement, quantum physics challenges us to reevaluate the fundamental principles that regulate the universe and the very essence of reality.

With the potential to reshape our understanding of the universe, the pursuit of a comprehensive and unified understanding of the quantum and relativistic domains continues to be one of the most exciting frontiers in contemporary science.

THE END